Solar System's Secrets

Color, and solve puzzles while you learn surprising facts about space

Nacionalidad El Mundo

Our solar system formed about **4.6** billion years ago. The four planets closest to the Sun: Mercury, Venus, Earth, and Mars are called the terrestrial planets because they have solid, rocky surfaces.

Two of the outer planets beyond the orbit of Mars — Jupiter and Saturn —. are known as gas giants; the more distant Uranus and Neptune are called ice giants

Sudoku Challenge

Enter numbers in the empty squares so that numbers 1-9 appear only once in each row, column and box

	8				7	6		2
6			3	4	1			9
	9	1	6					3
			8		3	1		4
				1	4	9		
1				6	9		5	8
4			1	3			9	
2	1	8				4	3	5
		6			2	7	8	1

The Sun

is the closest star to Earth, at distance from our planet 92.96 million miles This distance is known as an astronomical unit (abbreviated AU), and sets the scale for measuring distances all across the solar system.

SUN

Particles from the Sun can reach far beyond the planets, forming a giant bubble called the heliosphere.

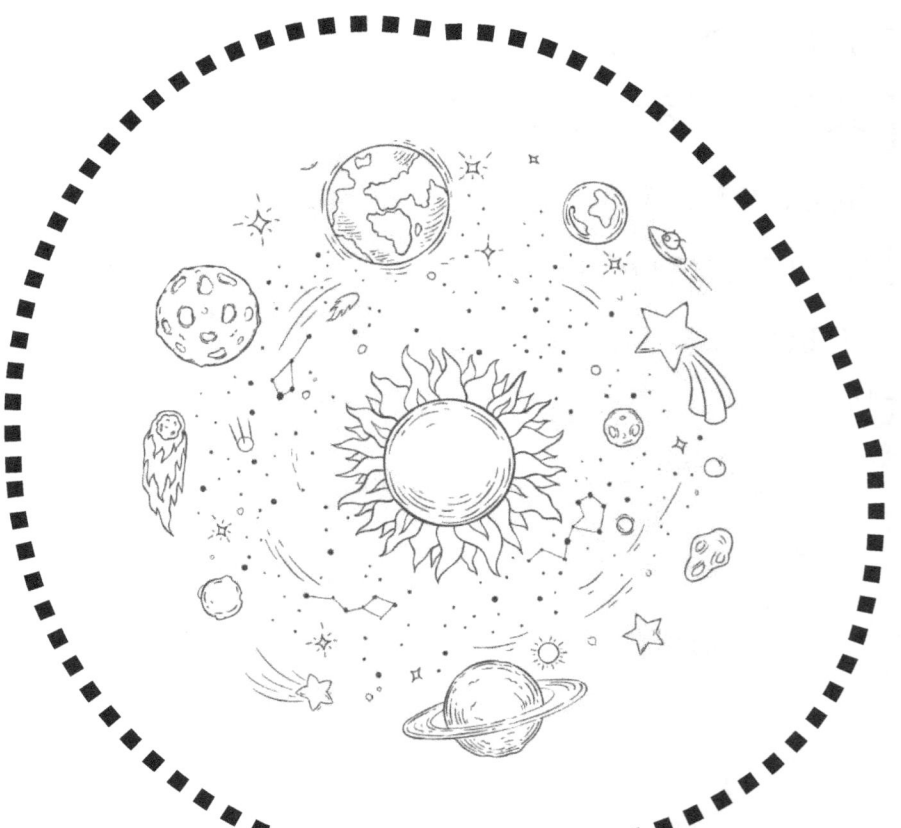

The Sun has six regions; the core, the radiative zone, and the convective zone in the interior; the visible surface (the photosphere); the chromosphere; and the outermost region, the corona.

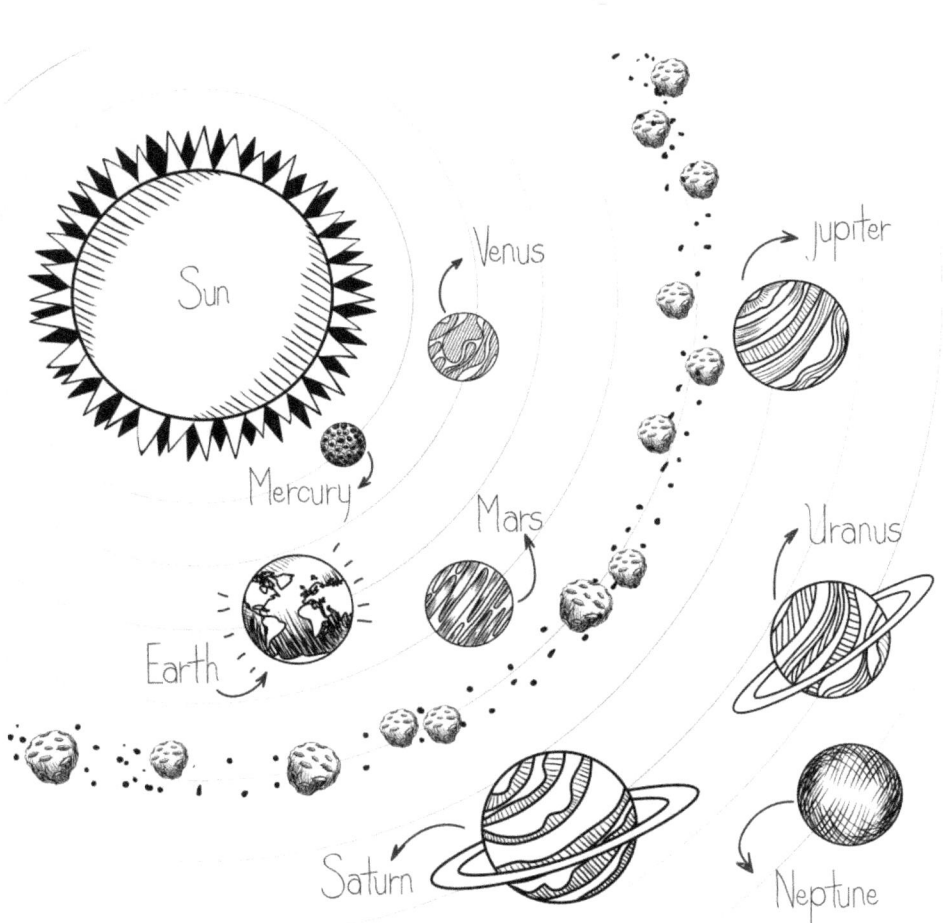

At the Sun core, the temperature is about **27** million degrees Fahrenheit, which is sufficient to sustain thermonuclear fusion. The energy produced in the core powers the Sun and produces essentially all the heat and light we receive on Earth.

Earth's atmosphere is primarily nitrogen and oxygen. Mercury has a very tenuous atmosphere, while Venus has a thick atmosphere of mainly carbon dioxide. Mars' carbon dioxide atmosphere is extremely thin. Jupiter and Saturn are composed mostly of hydrogen and helium, while Uranus and Neptune are composed mostly of water, ammonia, and methane, with icy mantles around their cores.

Mercury

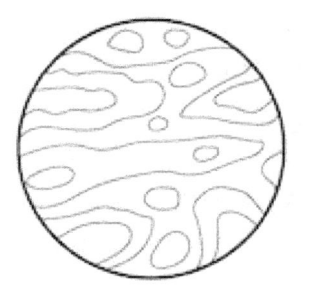
MERCURY

it's the closest planet to te Sun

Temperatures on Mercury's surface can reach 800 degrees Fahrenheit. Because the planet has no atmosphere to retain that heat, nighttime temperatures on the surface can drop to -290 degrees Fahrenheit.

Mercury travels at **31** miles per second, which makes it faster than any other planet, so it takes only **88** days to go around the Sun. On the other hand, its rotation is slow and it takes **175.97** days to complete a single revolution.

Ready for another Sudoku?

3	7				8		1	9
	5	9		1		7	3	8
2	1		7		9	6	4	5
	8	5	9			3		
9	2		3	7				
		1		8	5	4		
				9			6	7
		2			7	8		3
8	6							4

Venus

It's covered by a thick, rapidly spinning atmosphere, creating a scorched world with temperatures hot enough to melt lead

VENUS

Because of its proximity to Earth and the way its clouds reflect sunlight, Venus appears to be the brightest planet in the sky.

Now, invite a friend to play Dots and Boxes.

One places a vertical or horizontal line from one point to another and then the other and so on, the idea is to form little squares with these lines, but you can only place one line on each turn unless on your turn you are lucky enough to close a square, in that case, close the square, place your initial inside it so that it is known that it is yours and add another line in another couple of points, if you close a square again this time you will continue adding lines until is not possible and then you will add another line, but be careful because perhaps the line you add will help your friend to form many little squares

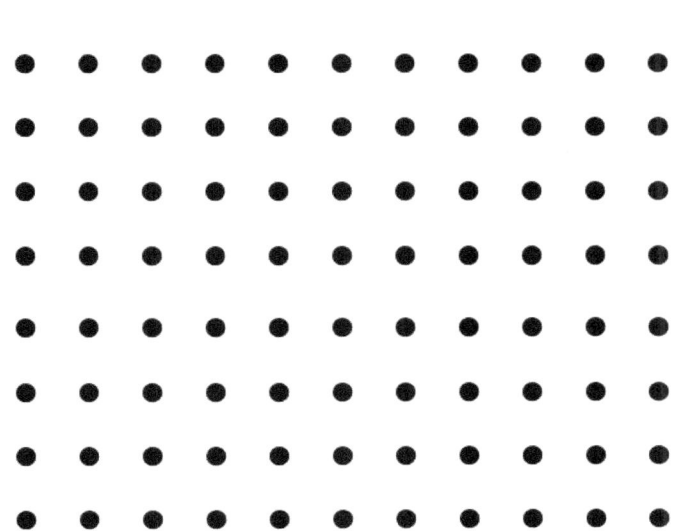

Venus rotates retrograde (east to west) compared with Earth's. Seen from Venus, the Sun would rise in the west and set in the east.

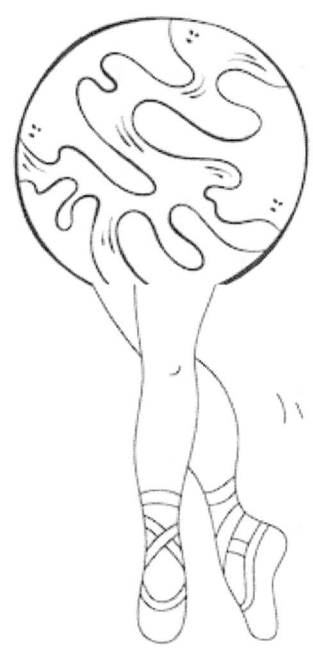

The Earth

The four seasons are a result of Earth's axis of rotation being tilted 23.45 degrees with respect to the plane of Earth's orbit around the Sun. During part of the year, the northern hemisphere is tilted toward the Sun and the southern hemisphere is tilted away, producing summer in the north and winter in the south. Six months later, the situation is reversed.

EARTH

Help Ralph to return to the Solar System

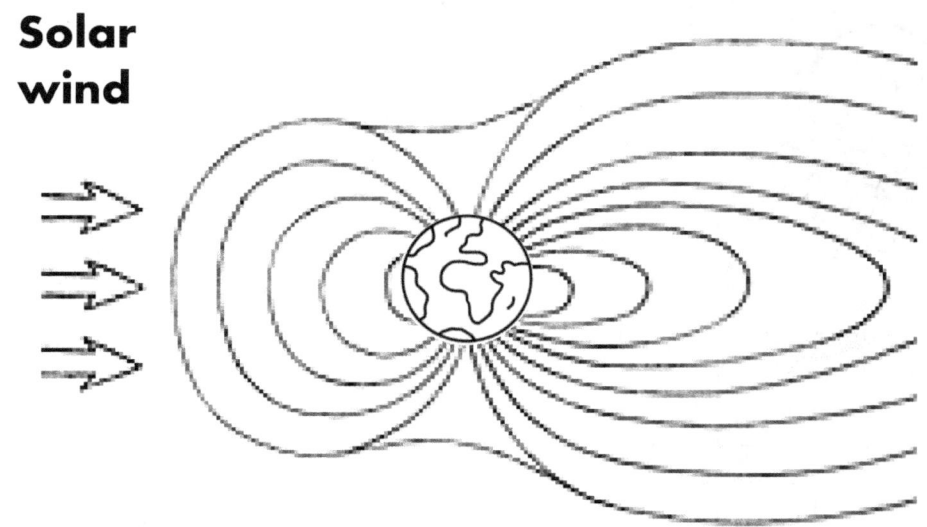

Most of the planets have magnetic fields that extend into space and form a magnetosphere around each planet. These magnetospheres rotate with the planet, sweeping charged particles with them.

Help Sally to return to Earth

Earth's lithosphere, which includes the crust (both continental and oceanic) and the upper mantle, is divided into huge plates that are constantly moving. For example, the North American plate moves west over the Pacific Ocean basin, roughly at a rate equal to the growth of our fingernails. Earthquakes result when plates grind past one another, ride up over one another, collide to make mountains.

New Sudokus

4		5	1	6	2	7		
				5			1	
1		7	4	9				6
3	4		7					5
2			5		9	6	4	
5	6		2				7	3
6	5	4			1	3		
8		2				9	5	
	9			2	5	4		

	4			5				
		5	9		8		4	2
8			4	7				6
5	3	6	2	9	7	4		
	1			8	4	6	9	
	9	8			6	7		5
	7					5	6	
1	8			6	5			4
6				2		8		7

The Moon

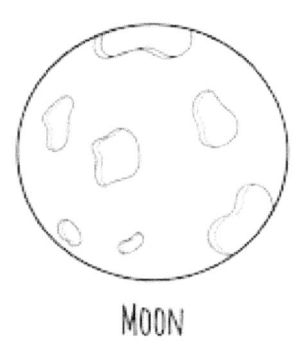

MOON

The light areas of the Moon are known as the highlands. The dark features, called maria (Latin for seas), are impact basins that were filled with lava between 4.2 and 1.2 billion years ago.

These light and dark areas represent rocks of different composition and ages, which provide evidence for how the early crust may have crystallized

Astronaut Neil Armstrong in **1969** was the first of 12 humans to walk on the surface of Earth's Moon.

The leading theory of the Moon's origin is that a Mars-sized body collided with Earth approximately 4.5 billion years ago, and the resulting debris from both Earth and the impactor accumulated to form our natural satellite.

NASA satellite observations help study and predict weather, drought, pollution, climate change, and many other phenomena that affect the environment, economy, and society.

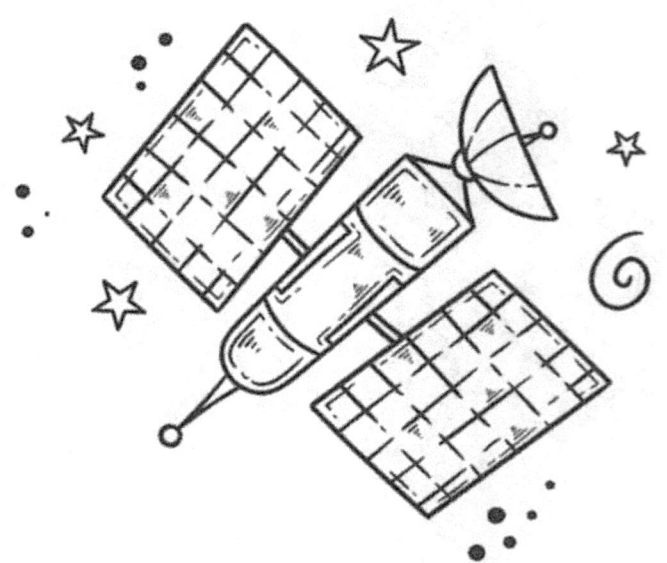

Mars

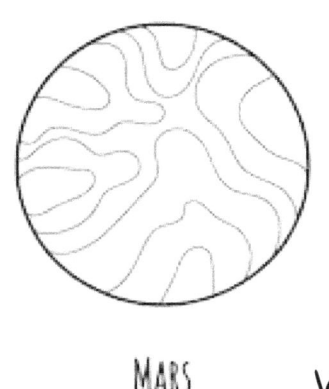
MARS

It's a rocky body about half the size of Earth. As with the other terrestrial planets volcanoes, impact craters, crustal movement, and atmospheric conditions such as dust storms have altered the surface of Mars.

The two moons of Mars are Phobos and Deimos. Phobos is slowly drawing closer to Mars, and could crash into Mars in **40** or **50** million years, or the planet's gravity might break Phobos apart, creating a thin ring around Mars.

Asteroids

Sometimes called minor planets, are rocky remnants left over from the early formation of the solar system about **4.6** billion years ago. Most of this ancient space rubble can be found orbiting the Sun between Mars and Jupiter within the main asteroid belt.

Find the members of the solar system

```
                        X A R
                      R A B A B J S T P
                    G Y R T W C I Z T M S L B
                  M T Q Z P Z A B O D L O D S P X Q
                Y G H U T X Q G D S P Q A B O V R P V
              N H A S B Z K T U W J F U A Y V A P J A C
              P L G M E G T Q T K L R D P I N O Z J J F
          Y H V O P P Y X E Q B S V I T R X R J P B I D
          F O V Q I W A Z Z I W N K O B I E L F B I G H
        Y G A D J Z C F A O K Z W H R T E N A L P I T D V
        A U J O G E M A C A J C P N E Y M T L K U R K A N
        R H R H B Y I E J J J R D I T M L X U P J S Q V J
    L H T R A E V U J T T L V X W S T K U D G Q H L J T E
    M H U E Y L S A I E E F Z J N A N O O M X O K F F Z C
    U E B W R J J E C X T V X H H D Q X R U N H O S T P L
      P T N Y S L B I V X I M G L W T J K H Q C S F P L
      C O E U X N T S Z M S K I G U T C A A K O Q K M Q
      P R K O S G A L X V P M Y P X B S M I G F O P E X
        F L H R V U T X F M E R E H P S O I L E H G Q
        W O U W I K J H T B H P J Z G E D E G M T I I
          D Q V D T S E K U M O N M O U I Y O Y G A
            F I J N U E Q K K V I Q P Z T T T G P C R
              X C K C F J T P I F S A E A R R C J
                H S A F R R V N B N Q U M C H O D
                  J K C Q H R U K E T L L I
                      H V O N Y O Z Q W
                            C D O
```

SUN	ASTEROID	MOON
PLANET	KITE	METEORITE
EARTH	HELIOSPHERE	

Shooting stars

Also are call as meteors, are bits of interplanetary material falling through Earth's atmosphere and heated to incandescence by friction. These objects are called meteoroids as they are hurtling through space, becoming meteors for the few seconds they streak across the sky and create glowing trails.

Help Tomy get into his spaceship

The largest known asteroid, Ceres, was discovered between Mars and Jupiter in **1801.** Originally classified as a planet, Ceres is now designated a dwarf planet

Sudoku time!

9								8
3	2			8	5	9		4
		1		9				
7	6	5	3	1	2		8	
2		3		4	9		5	7
4		8	5	7	6			
6						8		3
8		9	7			1	2	
	3			2	8		9	

			3	4	6			1
			7	9		8		
		9			8			7
3			8	6	7	9		2
	1				3	7		6
8			9			4	5	3
9		3				1	2	8
	8	1				6		
4	2		1	8	9	3		5

Jupiter

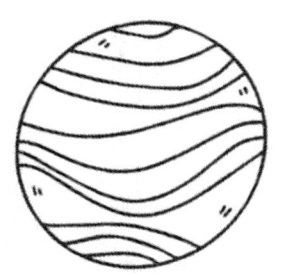

is the largest and most massive planet in our solar system, containing more than twice the amount of material of the other bodies orbiting our Sun combined.

JUPITER

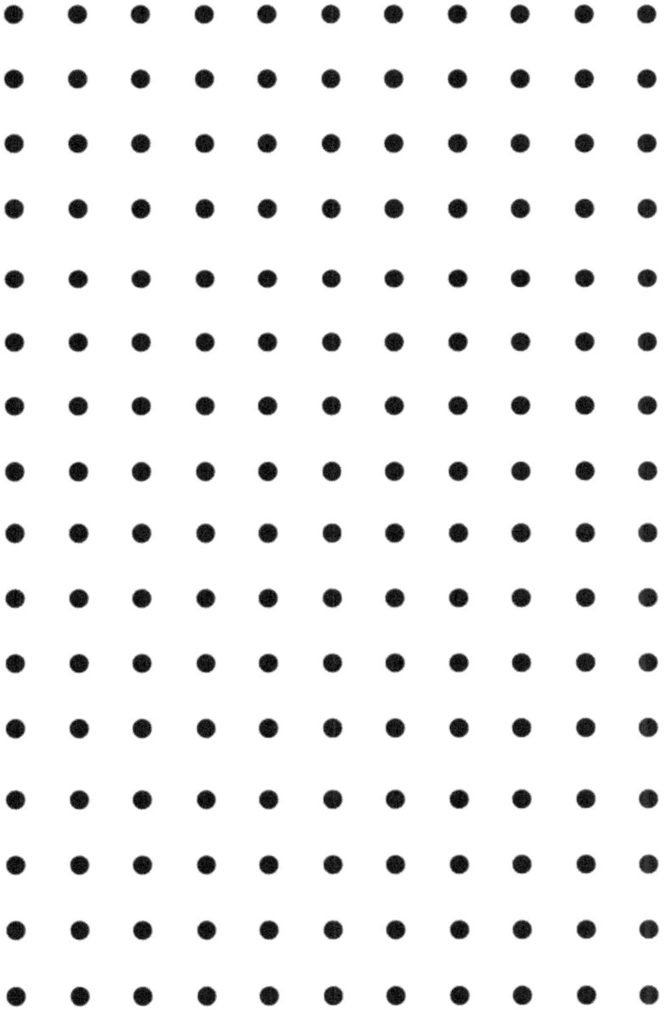

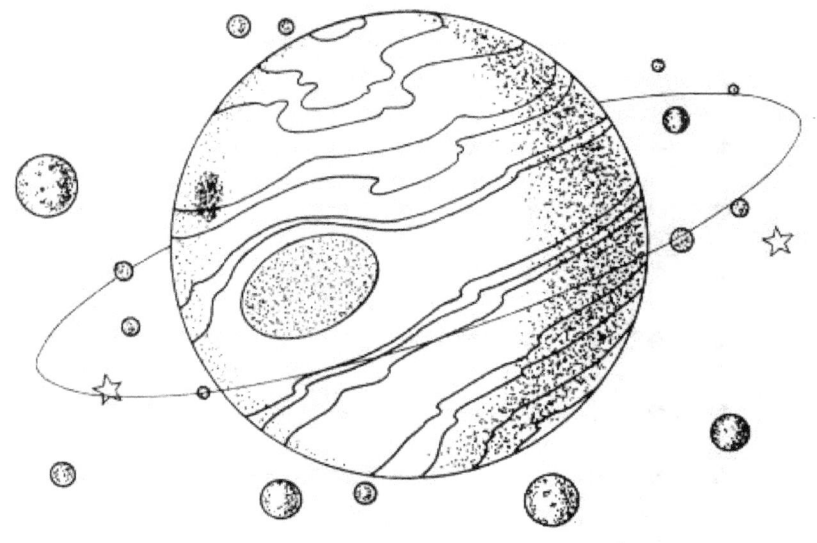

Jupiter has the largest moon in the solar system, Ganymede. Many of Jupiter's outer moons have highly elliptical orbits and orbit "backwards" (opposite to the spin of the planet).

Join the stars and discover what shapes the constellations have

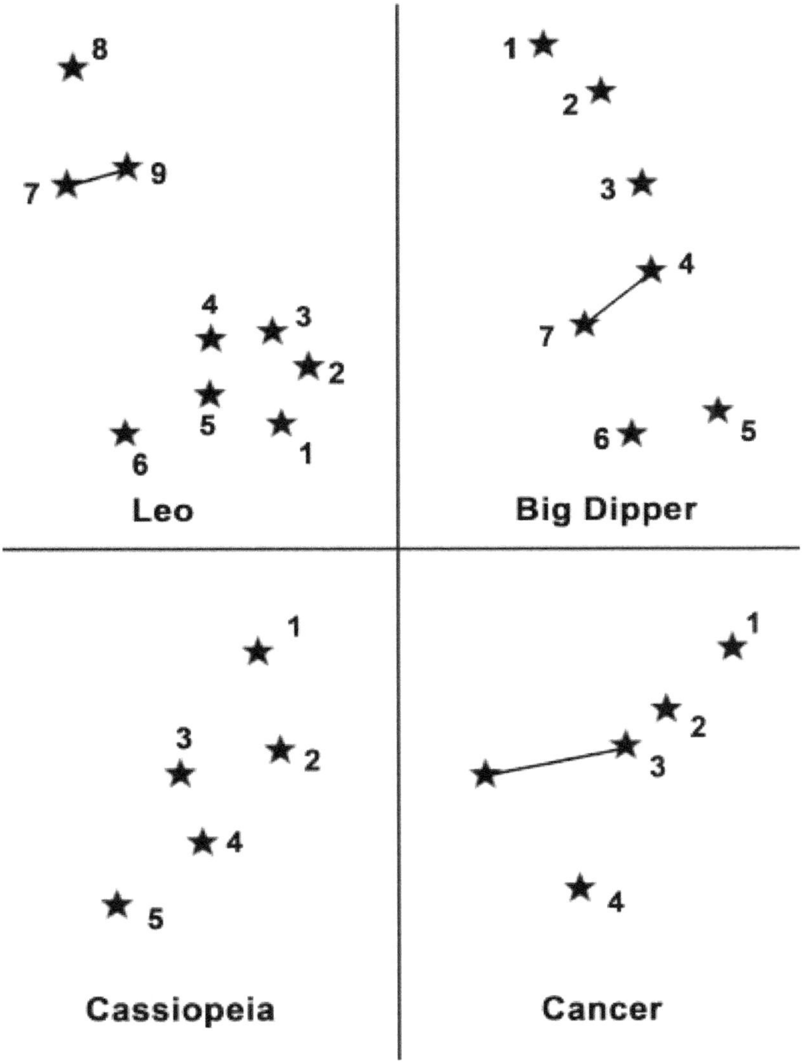
8
9
7
4
3
2
5
1
6
Leo
1
2
3
4
7
6
5
Big Dipper
1
3
2
4
5
Cassiopeia
1
2
3
4
Cancer

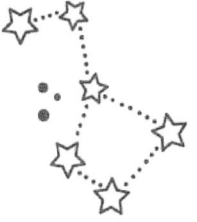

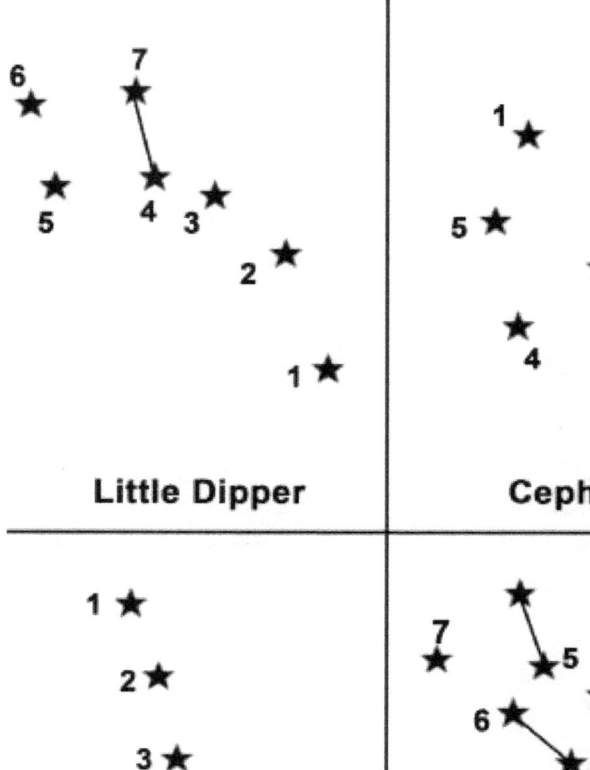

Saturn

It's made mostly of hydrogen and helium. Its volume is **755** times greater than that of Earth. Winds in the upper atmosphere reach **1,600** feet per second in the equatorial region. In contrast, the strongest hurricane-force winds on Earth top out at about **360** feet per second.

SATURN

Colorea de azul todas las naves espaciales

The chunks of ice and rock in Saturn's rings (and the particles in the rings of the other outer planets) are not considered moons, yet embedded in Saturn's rings are distinct moons or "moonlets." Small "shepherd" moons help keep the rings in line.

Enceladus a moon of Saturn, displays evidence of active ice volcanism

The great solar family

Find the members of the Solar System

1. elnEauscd
2. ntoirocamals nuti
3. hrEat
4. nurUas
5. cie sngtai
6. pshetohrpeo

(upside down answer key:)

1. elnEauscd — Enceladus
2. ntoirocamals nuti — astronomical unit
3. hrEat — Earth
4. nurUas — Uranus
5. cie sngtai — ice giants
6. pshetohrpeo — photosphere

Uranus

Like Venus, Uranus rotates east to west. Uranus' rotation axis is tilted almost parallel to its orbital plane, so Uranus appears to be rotating on its side. This situation may be the result of a collision with a planet-sized body early in the planet's history, which apparently radically changed Uranus' rotation. Because of Uranus' unusual orientation, the planet experiences extreme variations in sunlight during each **20**-year-long season.

URANUS

Discover our visiting friend

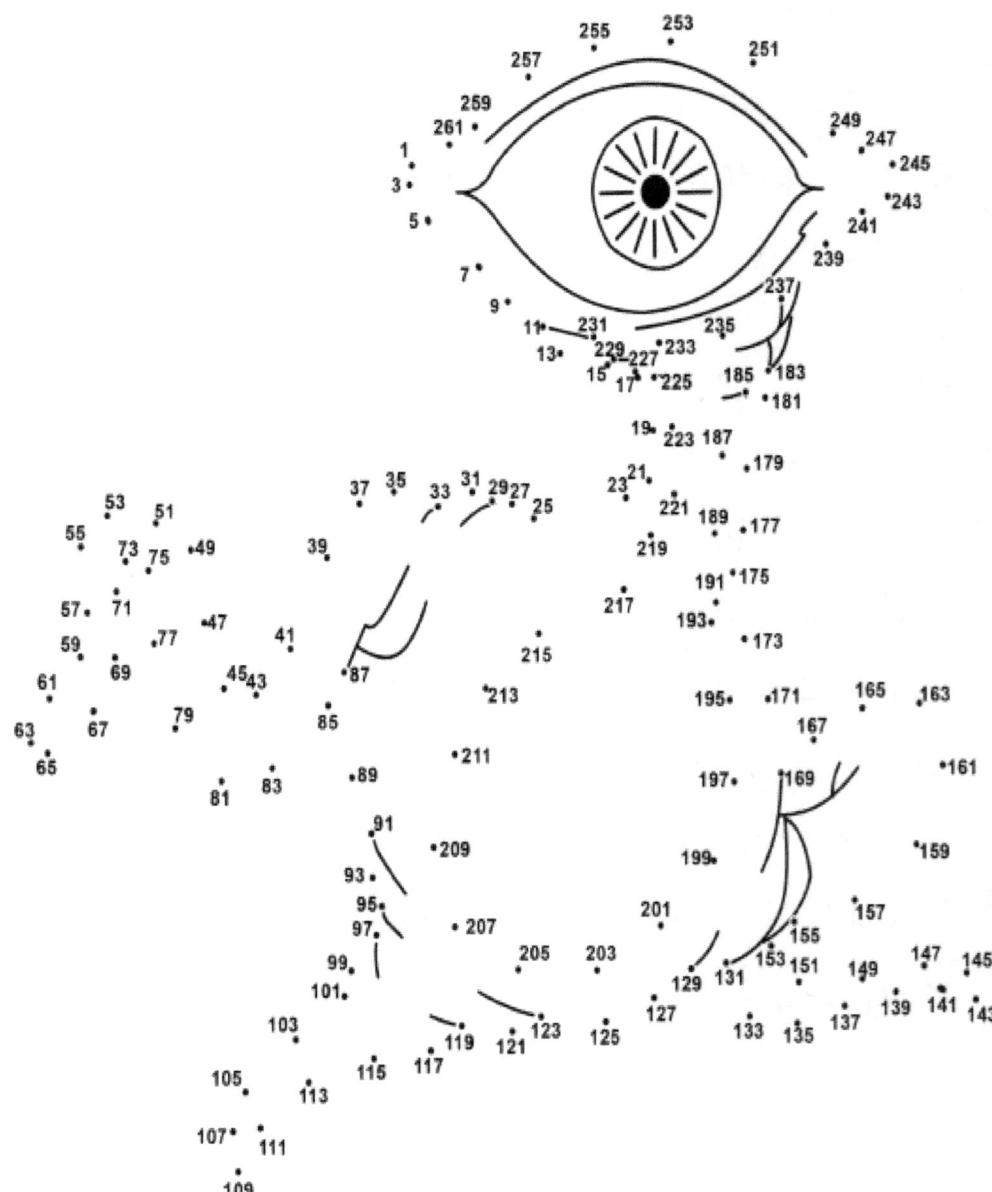

Help the astronaut solve these Sudokus This time only numbers **1-4** are needed

1			
2	4	3	
		2	
4	2		

1			3
			2
3			
2		3	4

2	1	4	
			4
4		2	1

		3	4
			2
3			1
4	1		

Uranus has **27** known moons. The inner moons appear to be about half water ice and half rock. Miranda is the most unusual; its chopped-up appearance shows the scars of impacts of large rocky bodies.

Discover the space explorer

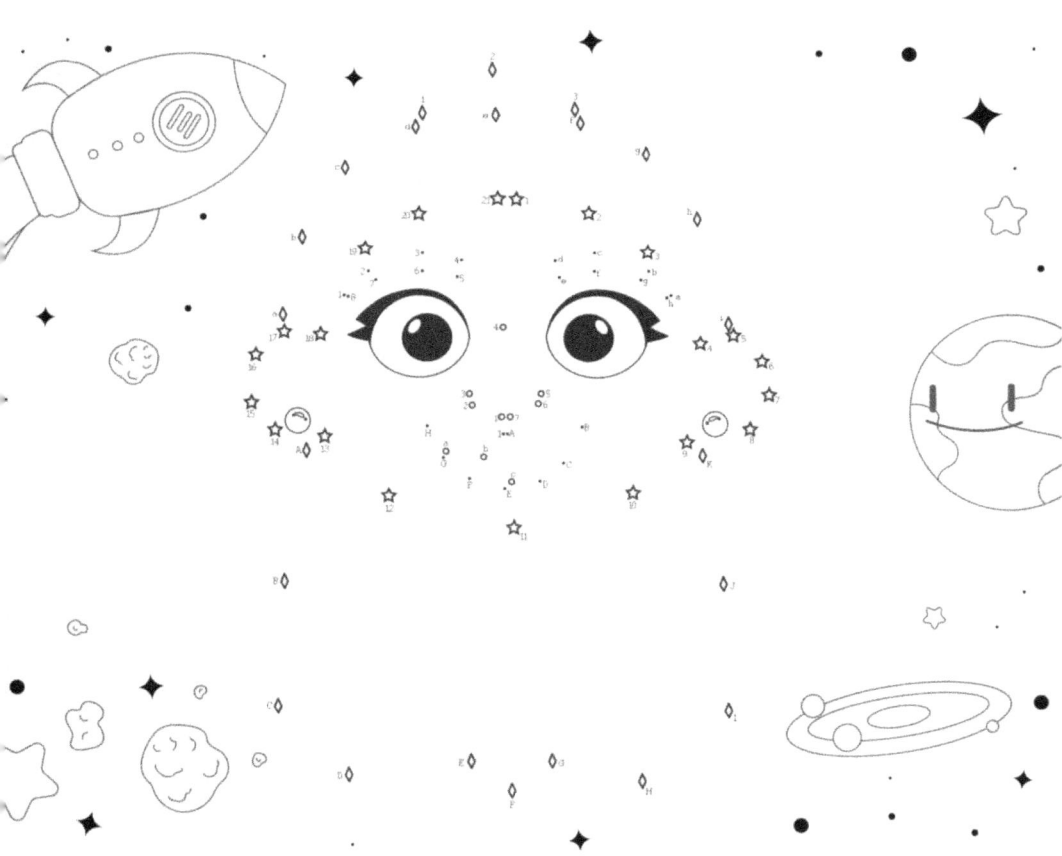

Neptune

Nearly 2.8 billion miles from the Sun, Neptune orbits the Sun once every 165 years. Interestingly, the highly eccentric orbit of the dwarf planet Pluto brings Pluto inside Neptune's orbit for a **20**-year period out of every **248** Earth years.

NEPTUNE

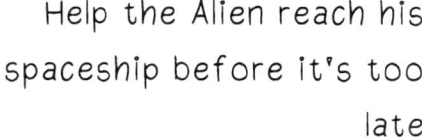

Help the Alien reach his spaceship before it's too late

Neptune's moon Triton is as big as the dwarf planet Pluto, and orbits backwards compared with Neptune's direction of rotation.

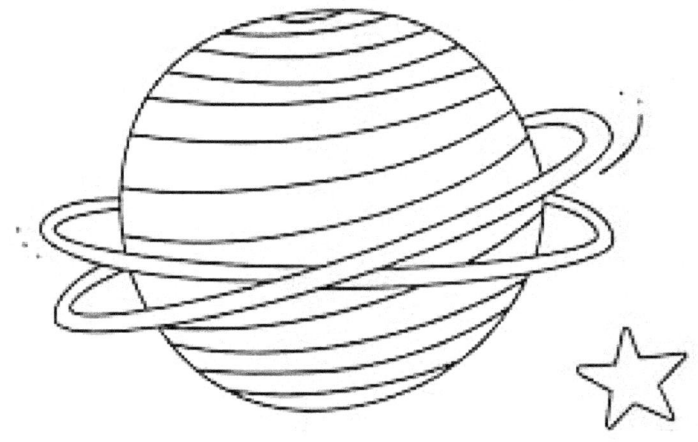

Join each planet with its Natural Satellite

1. _____ Earth   a. Triton
2. _____ Uranus   b. Miranda
3. _____ Mars   c. Titan
4. _____ Jupiter   d. Ganymede
5. _____ Saturn   e. Moon
6. _____ Neptune   f. Deimos

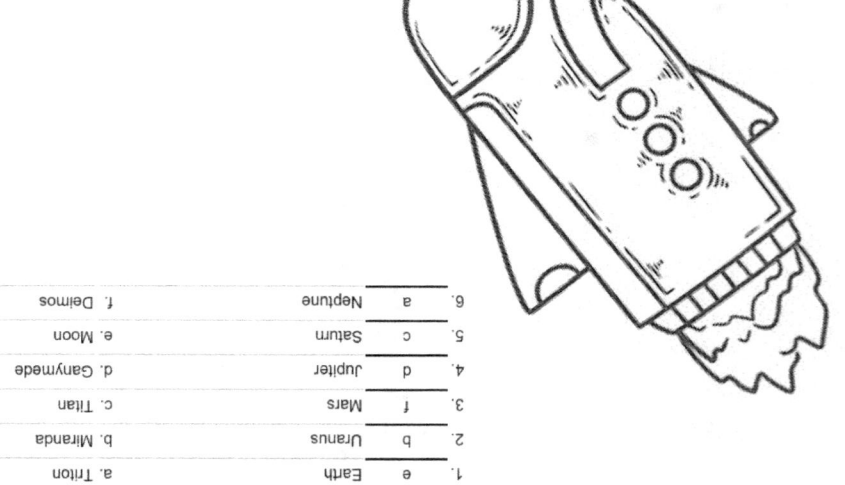

1. e Earth   a. Triton
2. b Uranus   b. Miranda
3. f Mars   c. Titan
4. d Jupiter d. Ganymede
5. c Saturn   e. Moon
6. a Neptune f. Deimos

Try these Sudokus

4		1	8	6	2			
2	7	8	3	5			6	1
		6		4		2		
	4			9	1	6		5
6		2					9	
1					6	3		2
		4			5	8		6
			6	3	8	5		
8		5			4		1	3

3		7	8			6		
		8			9	5		
5	6		2	1	3	8		7
4	7	6		2				
8	5	2	3		7			
			4	8			6	
6		4	7					5
			1	3	8	4		6
	3			4	6	2	8	

Researchers have found hundreds of extrasolar planets, or exoplanets, that reside outside our solar system; there may be billions of exoplanets in the Milky Way Galaxy alone.

Pluto

Is classified as a dwarf planet and is also a member of a group of objects that orbit in a disc-like zone beyond the orbit of Neptune called the Kuiper Belt.

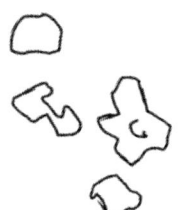

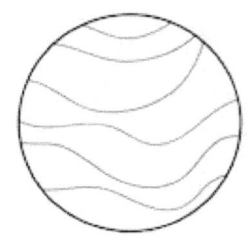

Pluto's 248-year-long elliptical orbit can take it as far as 49.3 AU from the Sun. From 1979 to 1999, Pluto was actually closer to the Sun than Neptune, and in 1989, Pluto came to within 29.7 AU of the Sun, providing rare opportunities to study this small, cold, distant world.

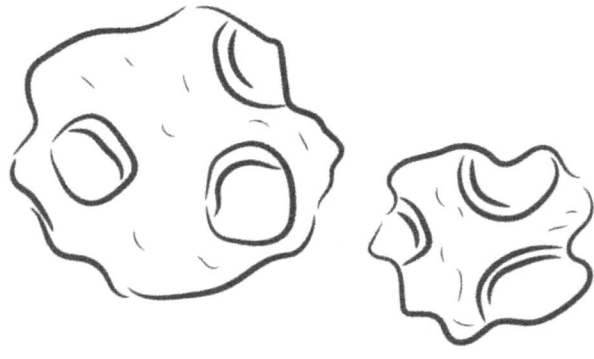

Pluto's very large moon, Charon, is almost half the size of Pluto. Charon is so big that the two are sometimes referred to as a double dwarf planet system.

Find Pluto's Moons

```
                                                    P K
X O Y C S M X I N A Q N V D R G F   E H
S S S L F H Y I I K E R B E R O S   C Q
L R                         M P     T C
K L     H Y D R A O H R E O D   K U N K
C Z     M D O T G Y Z D X Q R   N Q I F
D P     C I             P X     L Z D B
F U     W G   C N X Q T   N S   P B Q X
V R     H G   W T X I L   O W   C G W L
D M     Y R   J R   M E   E B   N Q L F
M X     R R   S R   T N   F X   H W B A
F Q     I X   X E         G A   W S F W
N C     O A   S R N S S T Y X   O H N U
J N     I I   Z C U J C C R Z   I I P F
H O     E L                     K J X O
W R     W G L L R I M R N C R F S G J Y
E A     N T C F E I G P E T U W W Q V W
N H                                 Y G
J C Y K Z C C C B Q B Z E U E M I M G Y
I W O D X F S P W G K R Z Q M N D X V L
```

CHARON NIX HYDRA
KERBEROS STYX

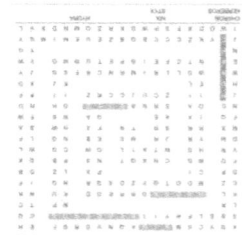

Comets are leftovers from the dawn of the solar system around 4.6 billion years ago, and consist mostly of ice coated with dark organic material

Each comet has a tiny frozen part, called a nucleus, often no larger than a few kilometers across. The nucleus contains icy chunks — frozen gases with bits of embedded dust. A comet warms up as it nears the Sun and develops an atmosphere, or coma. The Sun's heat causes the comet's ices to change to gases so the coma gets larger. The coma may extend hundreds of thousands of kilometers, forming a long, bright tail. Comets actually have two tails: a dust tail and an ion tail.

NASA's Stardust mission successfully flew within **147** miles of the nucleus of comet Wild **2** in January **2004**, collecting comet particles and interstellar dust. Within these samples, the amino acid glycine was found, which is essential for forming comets. living organisms

Find the Solar System Planets

```
                        M D
                        C Q
                      K Q H S
                      X M A J
                    L C E L W S
                    X N R T F U
J Q F B K V E E V C A X N K E A F F D G
H F V B S T A E N U T P E N S X L R M X
  M U J J R N B U R J V V S U N A R U
    C D T A E I R Y T U B R I V S G
        H I L K H C P P O M Z Z Y T
          J W Q Q T A G Q E J C H
        W C B S Q C K O N D S U N R
        V Y R L N C H P N R U T A S
      Y N E R E T I P U J B I J J P S
        I L A Z A O R   Y R U P P F K
    N M B A B T D       K F N D N K N
    W W B B E                 Z Q Y E L
  O K B N                         J E T C
  E V                                 W R
```

MERCURY   VENUS   EARTH
JUPITER   SATURN   URANUS
NEPTUNE

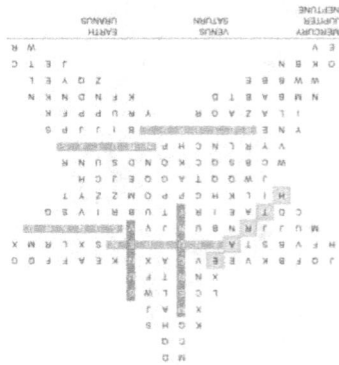

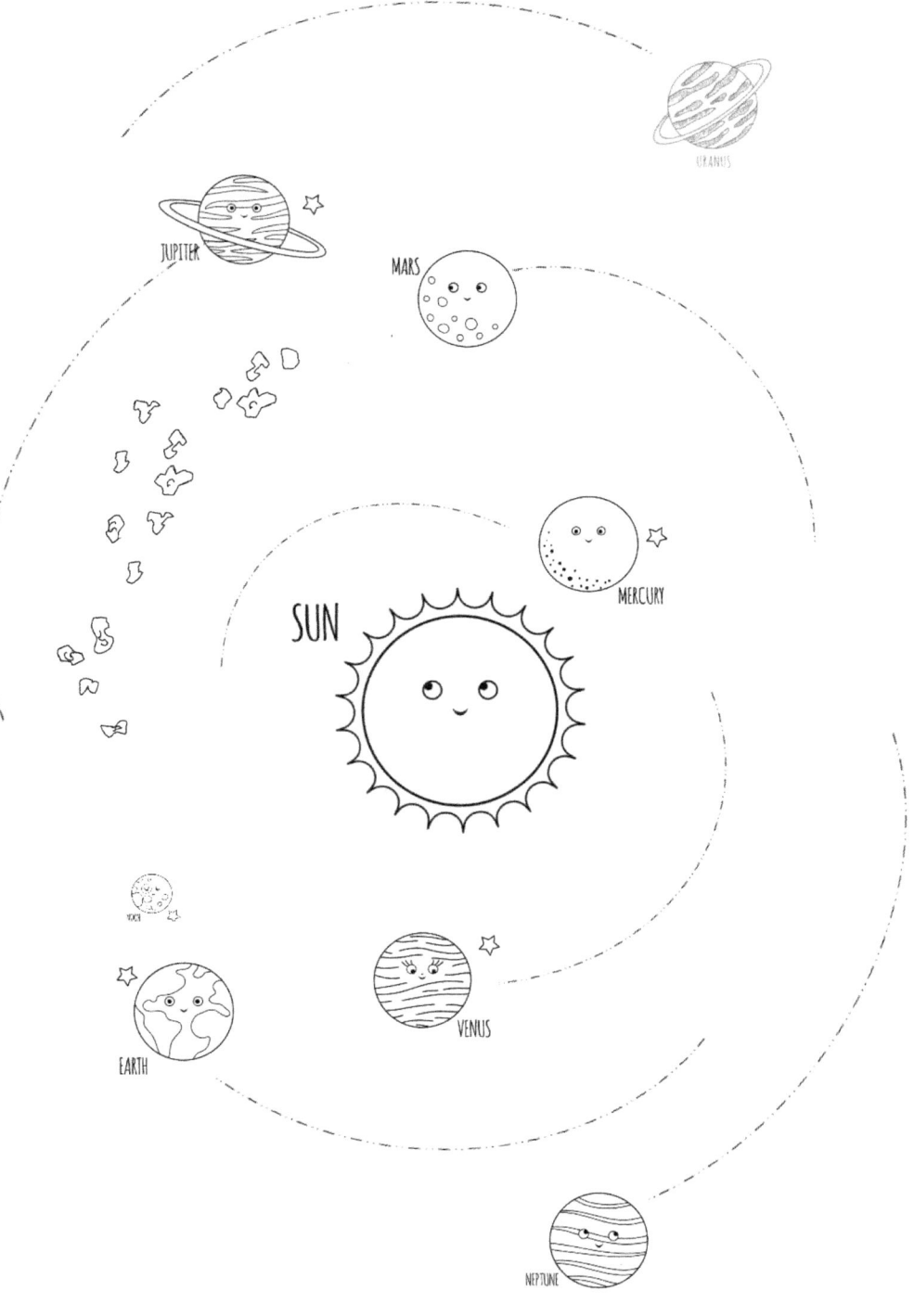

Bibliography

NASA. (s.f.). National Aeronautics and Space Administration. Obtenido de Solar_System_Lithograph_Set: https://www.nasa.gov/wp-content/uploads/2011/05/Solar_System_Lithograph_Set.pdf

Por que el Saber debe ser Universal

www.ingramcontent.com/pod-product-compliance
Lightning Source LLC
Chambersburg PA
CBHW070317230526
45470CB00002B/922